Life in Numbers

Our Favorites

Kristy Stark, M.A.Ed.

Publishing Credits

Rachelle Cracchiolo, M.S.Ed., *Publisher*
Conni Medina, M.A.Ed., *Managing Editor*
Nika Fabienke, Ed.D., *Series Developer*
June Kikuchi, *Content Director*
John Leach, *Assistant Editor*
Kevin Pham, *Graphic Designer*

Image Credits: All images from iStock and/or Shutterstock.

Library of Congress Cataloging-in-Publication Data

Names: Stark, Kristy, author.
Title: Life in numbers. Our favorites / Kristy Stark, M.A. Ed.
Other titles: Our favorites
Description: Huntington Beach, CA : Teacher Created Materials, [2018] | Audience: K to grade 3.
Identifiers: LCCN 2017029996 (print) | LCCN 2017039580 (ebook) | ISBN 9781425853228 (eBook) | ISBN 9781425849481 (pbk.)
Subjects: LCSH: Numbers, Natural--Juvenile literature. | Comparison (Philosophy)--Juvenile literature.
Classification: LCC QA141.3 (ebook) | LCC QA141.3 .S747 2018 (print) | DDC 513.2--dc23
LC record available at https://lccn.loc.gov/2017029996

Teacher Created Materials
5301 Oceanus Drive
Huntington Beach, CA 92649-1030
http://www.tcmpub.com

ISBN 978-1-4258-4948-1

We can choose a favorite.

Which pet?

He chooses
a dog.

Which color?

Four girls choose blue.

Six boys choose green.

Which treat?

They choose cake.

What is your favorite?